This guided study and journal can be used privately,
as a conversation guide for friends,
or in a small group setting.

This is also an accompaniment to the
Worship Design Studio series,
"Purveyors of Awe"
www.worshipdesignstudio.com/awe

May your awe bucket overflow.

For additional related content
and to purchase more copies, see
www.worshipdesignstudio.com/purveyors-of-awe

© 2025 Marcia McFee

Table of Contents

Awe: An Introduction	1
Beauty	11
Wonder	19
Meaning	27
Curiosity	35
Delight	43
Connection	51
Self-Giving	59
Dear Spiritual Leaders...	67
A Year of Awe	69
References	75
Author	79

Awe: An Introduction

I can remember a day when I was a young adult in my first year of living in New York City. I was about to graduate from New York University while working diligently toward my dream–a position with an internationally renowned dance company. It was evening and I had just come from a rehearsal for a production I was directing at my church. The light from the setting sun was streaming toward me on the sidewalk as I headed west toward my apartment. The song "Prepare Ye the Way of the Lord" from the musical Godspell, which we had just used in rehearsal and that had been a formative song in my upbringing, popped into my consciousness and I started singing it under my breath, walking lock-step with the rhythm of the song. Before I knew it I was in the midst of an awe-struck moment of feeling like anything was possible, I was in the right place at the right time, with the assurance that all was right with the world. It was a moment of certainty, of clarity, of hope, of joy at the possibilities that lie ahead.

Many years later (after actually achieving that dream-come-true dance career), I began in my PhD studies to discover the processes of the brain that occur when certain sensory-rich elements are in place. I learned that several things aligned that evening to create an overwhelming sense of awe, unity, assurance, and purpose in that moment: the beauty of the light, the connection of the meaning and memory of that song, the rhythm of my walking that sparked a neural slow-down of the part of my brain that differentiates the "self." I experienced what is called a "unitary state" in which I felt at one with everything–a sense of contentment and belonging that can literally add time to our lives. I'll never forget the feeling.

However, just because I can now tell you about the neural pathways that contributed to these moments of bliss, I am no less "wowed" by it. The very fact that our bodies are capable of such responses to the phenomenon of our world, the memories of our journeys, the meaning we make along the way, and the physiological rhythms we inhabit, is amazing and full of wonder. Research simply affirms what humans–especially poets and mystics who spend lifetimes musing about these things–have known all along: awe is good for us.

Awe is a profound emotional response that can significantly impact our well-being and perception of the world. Awe often arises from experiences that transcend our usual understanding, such as viewing vast landscapes, listening to powerful music, encountering extraordinary acts of kindness, or feeling heightened devotion and a divine presence connecting us all.

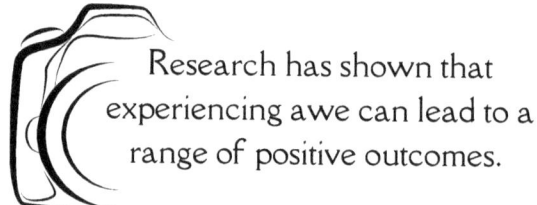

Research has shown that experiencing awe can lead to a range of positive outcomes.

Psychologically, it can enhance feelings of connectedness to others and the universe, promoting a sense of belonging. Studies indicate that individuals who frequently experience awe are more likely to report higher levels of life satisfaction.

Physiologically, awe has been linked to positive bodily reactions. When we experience awe, our bodies release hormones like oxytocin which can foster social bonds. This connection reveals how awe can serve as a natural antidote to the stresses of daily life, offering a momentary escape that encourages mindfulness and presence.

Sociologically and ethically, awe broadens our perspective, allowing us to see beyond our individual concerns. This expansive view can stimulate creativity and increase open-mindedness, encouraging individuals to approach challenges with greater resilience. In a world increasingly characterized by division and isolation, cultivating awe in our lives can be a powerful tool for fostering connection and empathy.

I posed a question to myself a few years ago about what might be "elements" for curating a life of spiritual depth. In this age of people identifying as "spiritual but not religious" or "nones and dones," and those seeking various ways to fill the hole left by a growing cynicism about organized religion, I have been curious about what a vocabulary would look like that incorporated modern research from the sciences about improving our spiritual wellbeing–which I understand to be an umbrella of mental, emotional, social, and physical wellbeing. I began to teach seminary courses related to this and students inspired me with their ideas and contributions to the inquiry. I began to compile a list of elements and to teach these in workshops and keynotes to glean reactions and gain anecdotal insights.

In recent years the publication of research related to the elements I was examining ramped up. These pages will offer a glimpse at some of this research. With the publication of Dacher Keltner's book called *Awe: The New Science of Everyday Wonder and How It Can Transform Your Life*, I became convinced that the overarching element that holds the whole list of elements of spiritual depth together is "awe." The ways we experience awe are not relegated solely to those unitary state moments that I described above. That was but one example of awe. Keltner lists eight aspects of awe that come from his research. While I don't match my elements with his, the fact that awe happens in many ways is enough for me to suggest that spiritual depth, like awe, can come via many starting points.

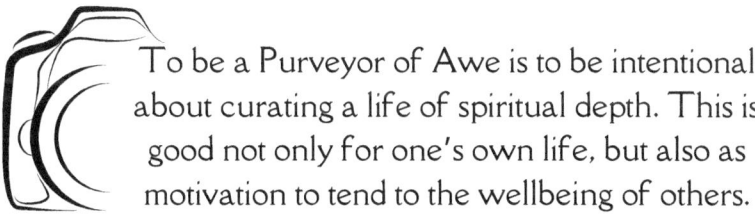

To be a Purveyor of Awe is to be intentional about curating a life of spiritual depth. This is good not only for one's own life, but also as motivation to tend to the wellbeing of others.

The connection of awe to the wellbeing of ourselves and also the way we *tend* to the wellbeing of others, is deeply important to me. I'm not interested in a purely "self-help" version of curating a life of spiritual depth–the world is too precarious and too many are at risk within it to ignore the ethical implications of our time and attention. The good news is that the research bears out that awe is not simply a feel-good emotion, it has consequences for how we perceive the world around us, how we inhabit the events that happen in our lives, and how we treat others. It can improve our life expectancy and our dedication to the good.

Purveyors of Awe offers an invitation to discover the difference we can make in our lives and in our communities if we engage in practices of awe. Indeed, I cannot think of a better mission. I have come to believe that one of the most important things that spiritual leaders and spiritual communities can do is be Purveyors of Awe for the greater community. A "purveyor" is someone who spreads or promotes an idea or view. The image of a camera–whether old school or the ones we talk, scroll, and text on–is the idea that when we examine life through the "lens" of awe and the elements of beauty, wonder, meaning, curiosity, delight, connection, and self-giving, we can make the world a better place. We can be about the business of inviting humanity to be deeply well and working to ensure that all have access to the possibility of wellbeing.

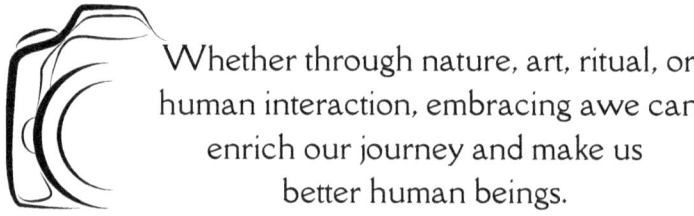

Whether through nature, art, ritual, or human interaction, embracing awe can enrich our journey and make us better human beings.

Being a Purveyor of Awe plays a vital role in emphasizing the need for individuals and communities to seek out awe-inspiring experiences in their daily lives. Philosopher and mystic, Simone Weill, said in her essay *Attention and Will,* "absolute unmixed attention is prayer." We are devoted to what we pay attention to. Or an analogy from my downhill skiing days, you really will go in the direction of your gaze so don't look at the trees, for God's sake! These days our attention is so scattered, so demanded, so co-opted that we give less and less time and attention to things that are life-giving. As a species, we are prone to attending to that which we dread and fear. It helped keep us alive. But now that the frequency of bad news is off the charts and it isn't keeping us alive, rather it is quietly sucking the life right out of us. We must engage in awe-filled practices that heighten our connection to that which is bigger than us, remind us that we are not alone, and offer us respite and resilience.

Learn. Reflect. Practice.

In this study journal you will be invited to move through three stages many times. It goes like this: you'll read something, then you'll write something, then you'll do something (or plan and prepare to do it). This Learn/Reflect/Practice rhythm is a way for you to process through various "formation" channels: Information. Exformation. Formation.

We are used to learning through taking in "information." That's why you pick up a book–especially a study book–to take in an idea.

The next step, however, is when we begin to really work with it. We "exform" it. In other words, we allow the information to be a prompt as we journal about how the information shows up in our lives externally in experiences.

And then we take that into the future through "practices" which will increase the durability of the retention and incorporate it further into our lives by creating an embodied experience in which the concept gets to blossom. Thus begins the "formation" of our future selves.

So let's give it a whirl. I just gave you an example of a moment in my life when I felt "awe-struck" and described the circumstances and the elements that I believe led to this amazing feeling of at-one-ness (the place, the light, the song, the rhythm, the walk). Now it's your turn. Take some time on the next couple of pages to remember and reflect. Then we are going to figure out a practice from the reflection.

And here's the thing: when I give you a prompt, don't feel bad if you can't figure out how to respond immediately. Simply sit with it and if nothing comes, move on. Or try drawing and doodling. Because the brain will keep working on it and later, when you least expect it, you'll come up with something and return to the page. The other thing to do when you aren't sure how to address the prompt is to write whatever the heck you want. Because that's often the way "aha's" work–kind of sideways. Just let whatever comes to you flow out and who cares if it is "right?" That's the thing about "exformation." The only thing that can exform is what is inside of you. So it is yours.

Try to recall an experience of being "awe-struck." What were the circumstances? What were the elements at play that led you to this heightened sense? How did it make you feel?

*I, through the abundance of your steadfast love,
will enter your house;
I will bow down toward your holy temple
in awe of you.
– Psalm 5*

Awe Practice Example

OK... the "do something" part for each of our elements of spiritual depth could go something like this:

Since sharing my New York City awe-struck experience, I've thought about how I might curate circumstances that might put me more frequently in the possibility of awe. For some reason, I just don't do earphones very frequently. I love music. And I know that listening to music has a huge effect on me–my mood, my energy, my disposition–all factors that make me a better human being with a more positive outlook on life. But I also need silence or white noise when I'm working because music is so powerful that anything that's playing will pull me in to the point of distraction. Then I forget about re-engaging the music when I take a break and I get run down faster and less able to offer my energy to others. So telling you the story has reminded me to curate more intentional times of walking with music. So here is my list:

- get better earphones that are comfortable and easy to put on
- make playlists of songs that are meaningful to me
- take walks during the "golden hour" before sunset
- keep time with the music

It sounds like an exercise program. It kind of is, except that it is exercise for my awe muscles. And like all exercise, it is possible that it might take me a minute to actually get down to it. I mean right now it is a frigid tundra where I live and walking to the beat of music could end up with me on my butt. That's ok. Moving through the exercise of figuring out what could offer me more awe-exposure is helpful in that it plants an idea in my mind and therefore my "noticer" has gone up already. I'll notice what I'm doing at twilight. Am I still sitting at my computer trying to get just one more thing done on my to-do list? Am I taking a moment to appreciate something–anything–that will help me get a perspective about what is real for me in this moment, right where I am instead of moving to the evening news immediately or doomscrolling the latest outrage? Simply turning up our "noticer" is a really good start. Imagine in the next page a practice for yourself based on your own experience of what lights you up, whether you think you'll get to it right away or not.

Awe Practice: Memory to Experience

What are you beginning to come up with for curating the possibility of more awe in your life based on your memories of experiences of awe? When have you said to yourself, "I need to get more of THAT in my life!" Start a list here and come back to write more things down as you move through this study. Get specific about what you might need to make it happen.

Simply being in a context of awe leads to a "small self." We can quiet that nagging voice of the interfering neurotic simply by locating ourselves in contexts of more awe. – Dacher Keltner, Awe: The New Science of Everyday Wonder and How It Can Transform Your Life

Beauty

While "beautiful" itself is an adjective, there are many ways to describe our experience of beauty. We might say something is "stunningly" beautiful. Or "breathtakingly" beautiful. Or, my favorite, "achingly" beautiful. What we see in common with all these ways of trying to express the awe at what our senses have perceived is "aesthetic arrest," a term popularized by Joseph Campbell with a long history from the philosophy of beauty. We have a kinetic response. Our bodies literally suck in breath and hold it for a moment. We may stop moving in our stunned state. We perhaps feel a physical ache of yearning to capture the import of the moment. There are often guttural responses of "ahhhh" or exclamatory bursts of "wow" and "oh my." We are quite literally "stopped in our tracks." We realize we are faced with an experience in which time must slow down in order to "take it in" and solidify the image in our memories so the effects might live longer within us. We want to share in the radiance and the resonance of it. We want to live in the awe of it for a time.

Listen to the Hebrew poet's yearning to dwell in this state of being:

> **One thing I ask of you, YHWH,**
> **one thing I seek:**
> **that I may dwell in your house**
> **all the days of my life,**
> **to gaze on your beauty**
> **and to meditate in your Temple.**
> – Psalm 27 (Inclusive Bible)

In the summer of 2018, I spent several weeks wandering through Ireland on my own. I had the experience of "aesthetic arrest" in a valley just south of Westport on the Wild Atlantic Way. It was mid-morning and I had to get to my next bed and breakfast about two hours away with several stops along the way, including the city of Galway with so much to see there. Yet once I rounded a corner of the road which revealed the valley before me, I knew I had to stop. To dwell. To meditate in the "temple" of this place.

As was the theme of my sojourn that summer, I threw the plans for the day out the window, so to speak, and decided I would stay as long as my soul needed to stay. There was no rationale for the change in plans and even now I have a hard time explaining in words why I was drawn to it–or more to the point "arrested" by it in that moment. I was "beguiled by beauty," as my friend Wendy Farley describes in her book by the same name.

The feeling I had that day was different from the awe of the ecstatic walking to the beat of a piece of music I described in the introduction. This reaction was quite the opposite kinesthetically. I was so moved that I couldn't move. Or at least couldn't move quickly away from the scene, rather I was lured into what I now call the spiritual practice of "lingering longer." To linger is an act of resistance in our fast-paced world of spending time living "productively." What a feat it is to actually embrace a change of plan because we've encountered a moment of such value that we simply must not miss it.

This is one of the healing aspects of beauty. It insists that we get off the train we are on in order to dwell, to linger, to gaze, to meditate. And just like the Psalmist, we can ask for this, seek this, because we know we need its healing.

 Beauty is sacred in the sense that it insists we be arrested into a "setting apart" of time and space to dwell, to linger, to gaze, to meditate.

The Ireland trip I'm describing was a prescription from my therapist. She had abided with me during a stretch of several difficult years. I was drained. Joy, energy, motivation–had run out. Who I was, what I felt, what I knew to be true, and what I wanted out of life had pretty much taken a back seat to simply surviving, to getting through. Oh, I was very high-functioning and many who only know me professionally might be surprised by this. I had gotten very good at compartmentalizing and putting on a good face. But the full reality was very different.

And so the prescription from my therapist went something like this: "When was the last time you stopped to enjoy life?" Ugh. Not what an Enneagram Three wants to be asked. Not what a person who has built a protective wall of productivity around themselves wants to be asked.

"Uh, I don't know."

"It's time, you know."

"I know." Sigh. And then the flood of tears that comes when a truth has been spoken into your life.

Perhaps part of the preparation for the spiritual awe practice of lingering longer in the face of beauty is the realization that we need it desperately. Without the ability to be stopped in our tracks and redirected, if only for a few moments or a few hours, we run the risk of bulldozing the beauty right out of our lives, leaving leveled tracks of compressed dirt that stretch on forever.

And so my therapist asked me to make a date later in the year, choose a place that I had promised myself I would someday visit, and most importantly, not over-plan it (she knew me well). It was an invitation to be open to whatever presented itself in the moment. She knew what could happen when places of unknown-to-you beauty and an openness of spirit converge. You don't have to go to a country an ocean away, although I would never discourage it if you can do it, to combine these elements of beauty (it's everywhere), discovery (be open to surprise), and a willingness to linger longer (keep the time frame loose, if at all possible). Take some time and space to reflect on how beauty affects you, what keeps you from getting more of it, and make some plans.

What is your experience of being "stopped in your tracks" by beauty? Describe the scene, describe the feeling. What effect could it have on your life to get to practice "lingering longer?"

We live between the act of awakening and the act of surrender. Each morning we awaken to the light and the invitation to a new day in the world of time; each night we surrender to the dark to be taken to play in the world of dreams where time is no more. At birth we are awakened and emerged to become visible in the world. At death we will surrender again to the dark to become invisible. Awakening and surrender: they frame each day and each life; between them the journey where anything can happen, the beauty and the frailty.
– *John O'Donohue, Beauty, The Invisible Embrace*

The story of that Irish valley and the beauty there is not finished. There is another aspect that blew my mind into an even deeper experience of awe. And it is one that Dacher Keltner names as one of the eight aspects of awe: moral beauty.

The new friends I had just met in Westport had suggested traveling the road through the valley and told me to watch for a marker by the side of the road. When I found it, the inscription told the story of the events of 1847 during the Great Famine. Around 400 poor and starving people had walked eleven miles upon hearing that authorities would be lunching at a lodge in the next town. They set off to beg for assistance and for food. This was to no avail and many of them starved along the way to and fro, their bodies lining this very road through one of the most beautiful places on earth. There was a pile of stones around the marker. No instructions about what to do, but I felt an innate urge to pick one up.

My friends had told me there would be another marker down the road as well and so I set off to find that one, trying to wrestle with the juxtaposition of immobilizing soul-healing beauty and the horrifying story of this place that was beginning to unfold.

At the next marker, I saw the other pile of stones and I read the story of what happens each year along this road. A commemorative walk of these eleven miles takes place and people carry a stone, perhaps a symbol of the weight of genocide that is part of our human story. Irish people lost a quarter of the population during the famine to starvation, disease, and immigration. And so the people walk to remember and to continue the healing in this beautiful valley. But they do not walk alone. After hearing about this story of terror on this road, the Choctaw Nation sent money to the Irish people during the famine, spurred by their own experience of devastation on the Trail of Tears. Choctaw representatives come to walk in this annual remembrance, and others such as Desmond Tutu and, most recently, representatives from Gaza.

I got chills as the story unfolded. And after hearing Keltner talk about "moral beauty," I now know why. Just as powerful as the initial awe I felt in that beautiful place were two emotions: horror, yes, at the atrocities, but also deep awe at the story of response and solidarity from people around the world, the resilience of the Irish people in the generations that followed the famine, and the courage of all those who withstand great suffering and inhumanity, discrimination and injustice.

Since my trip in 2018, I now take groups of women on annual pilgrimages to Ireland. The trips are called "Creator, Badass, and Saint," helping us to reflect on our own creativity, advocacy, and legacy in this world. I take the women to this road. We pick up stones and walk from one road marker to the other (that stretch is shorter than eleven miles, but long enough). We walk in silence, feeling the weight and sharp edges of the shale stone, praying for all who suffer and for our own courage to be actors of moral beauty in this world. No matter the weather (and we've seen everything from sun to gale-force winds and rain), it is deeply beautiful and, to a person, we are awed. It is perhaps one of the most memorable parts of the trip for many.

> We were made from dust and to dust we shall return. We open ourselves to the awe that is the precious beauty of life in between the first and last breath.

As I mentioned earlier, my friend Wendy Farley writes that beauty undergirds a theology of goodness and survival. She says, "All things are beautiful–not by a standard of 'pretty' as seen by our eyes but by an essence of sacred worth that is seen with the spirit. This is the root and heart of compassion and justice. Beauty is the threshold to Divine Goodness and a door into radical compassion. The difficulty and crisis of the world is overwhelming. It is virtually impossible to bear it without very deep resources."

Contemplative practices, she says, are the way we intensify our spiritual capacity to withstand and stand up to cruelty in the world. The scientist and the theologian deeply agree. We need more exposure to the awe of beauty, not just to fill our own souls but to increase our capacity for compassion and to bolster our courage.

And so, as Purveyors of Awe, can we advocate in our lives, our families, our communities for making set apart time and space to deepen our practices of awe through beauty? We too often forget that this is not optional for our wellbeing, it is essential.

Awe Practice: Nine Beautiful Things

Psychologist René T. Proyer studied the effects of recalling and noticing beauty each day. His "Nine Beautiful Things" Practice goes like this:

Set about 15 minutes before going to bed to think about nine beautiful things that happened during the day, three each in the following categories:
1. Write down three beautiful things on human behavior (morally, positively valued behavior, ie good deeds).
2. Write down three things you experienced as beautiful in nature and/or the environment.
3. Write down three beautiful things in general that you noticed during the day (referring to aesthetics, like art, music, architecture, etc).
4. Note why you found each of these nine things beautiful.

When we naturally notice and appreciate the beauty around us, it can lead to all kinds of different benefits. Studies show it can make us feel more satisfied in life, have a stronger sense of meaning, and act with more kindness towards others and the environment. – Shuka Kalantari, about the research of Rene Proyer,"Why We Should Seek Beauty in the Everyday Life" (The Science of Happiness Podcast), Greater Good Magazine

Wonder

Rabbi and mystic, Abraham Joshua Heschel, tells us that "to be spiritual is to be amazed." He speaks not of momentary bouts of amazement, but that "everything is phenomenal; everything is incredible; never treat life casually." This is not a *surprise state* in us when faced with something extraordinary, but rather a *state of being* that allows us to see the wonder-full in the ordinary. This reminds me of one of my favorite Charles Wesley hymns, "Love Divine, All Love's Excelling," in which the poet yearns to become a "humble dwelling" for the "joy of heaven to earth come down." The hymn ends with a phrase, a prayer, that we be "lost in wonder, love, and praise."

How we get "lost in wonder" differs from one person to the next. Our brains are simply wired differently. I have used the Multiple Intelligences Theory by cognitive scientist, Howard Gardner, for years as a way of talking about our sensorial diversity. Some of us are "tuned in" most easily to words, some to music, some to visuals and space, some to movement, among others. What "moves us to wonder" is connected to how facile the various parts of our brains are based on our experience and practice.

Visuals, movement, and music are my go-to "wonderbringer" senses. I cowrote a book about skiing and snow-boarding as spiritual practice when I lived in the Sierra Nevada mountains near Lake Tahoe. Those years of being on my skis in the stunning environment of that place was a magical feast for my senses. One night while writing the book, I got the rare opportunity to ride in the Snow Cat machine that creates those delicious corduroy runs I adored. The driver, Sarah, was a concert pianist besides being a "snow farmer" and she loved to plow while listening to movie soundtracks. At the beginning of my time with her that night, I

was truly in awe–both terrified because they tether the Snow Cat to a tree at the top of a black diamond run so it won't flip end over end down the mountain, but also absolutely mesmerized by the motion of the machine, the music soundtrack that made it feel incredibly dramatic, and the visuals of slopes I had skied many times now filled with the glowing night eyes of the wildlife. But after like... uh... four hours of it (with four more to go), my wonder gave way under the sheer repetition.

Have you ever thought or said, "the novelty has worn off?" It turns out our brains are wired in such a way that we return to a prescribed "baseline" after new, novel, or positive encounters happen in our lives. Something that once felt extraordinary shifts over into the "ordinary" category in our perception catalog. We "get used to it" and we stop seeing it vividly in our memories and imaginations. The upside is that it makes more space in our brain functioning for other information, other phenomenon, to be processed and it increases our ability to handle the unexpected with more energy. Our brains are quite efficient in this way. This shifting has been termed "hedonic adaptation." The other upside of this ability is that this is true for negative experiences as well. We can eventually return to a baseline of contentment or happiness in which the negative no longer holds our immediate attention all the time.

However, the downside of hedonic adaptation is that we stop feeling the "wonder." Most of what we experience repeatedly becomes commonplace or within our comfort zone. We stop noticing with wonder that the sun comes up each day or marvel that domesticated animals love having us as part of their "pack"–all stuff that is actually pretty amazing. We start to think that there's nothing new under the sun and begin to see our world as full of scarcity rather than abundance. Our "first glance stance," as I call it, gets dulled or even cynical.

What would it be like if we could live as Rabbi Heschel described–"taking nothing for granted?" It takes effort. It requires a waking up of the brain and the suspension of "business as usual" processing. It takes practice.

But the payoff is big. If we come at life with wonder rather than cynicism as a "first response" it is possible we might notice the amazing abundance, amazing grace, amazing transformation all around us. It strengthens the synopses that tether the everyday to our sense of the sacred. We might even wonder at the miracle of our own being.

> **For all these mysteries I thank you—
> for the wonder of myself,
> for the wonder of your works—
> my soul knows it well.**
> – Psalm 139 (Inclusive Bible)

The Hebrew poet describes the recognition of wonders as a "soul" kind of knowing. The tricky thing about finding words to describe awe and wonder is that it is an emotion, an experience, that often defies words. It is a feeling and knowing that sometimes just *is*, rather than something we can describe or translate to someone else. Monica C. Parker, in her book, *The Power of Wonder,* says that understanding wonder is "part science, part soul." She says that some things that are unexplained might be explained tomorrow. Likewise what we think we understand could be disproved at any time. But the adventure of exploration has value. "We don't have to understand wonder to experience it." We suspend judgment and expectation in order to invite a kind of "poetic faith" in the goodness of the journey itself.

Some translations of the Psalm poem above say, "I give you thanks that I am fearfully, wonderfully made." Sometimes our perception of ourselves can be fraught with thoughts of fear. "Why the heck did I do that? What is wrong with me?" We call ourselves all kinds of names we would never call someone else. Actually, the Hebrew word used for "fearfully" in that passage is better translated as "awesomely" made. Awe can include both amazement and fear at the enormity, the vastness, of what we are experiencing. We often feel like we ourselves are a vast project with too many "bugs" in the system. But stop and actually think about the "system" itself. Talk about amazing. Awesome. Full of wonder. Each breath. Each heartbeat. Every little thing that makes us human. It is nothing short of miraculous. You are a wonder. It is a wonder that any of us exist. The probability of our existence in relationship to the vastness of what we know of the universe at this moment (which turns out to be simply a fraction of the whole) is infinitesimal. Take a moment. Take it in. Then say, "I exist. Wow."

Just because modern science can explain how the system works, doesn't make it any less of a wonder that we are put together in just this way.

What is something that was once a "wonderbringer" for you and now seems quite ordinary? Reflect on ordinary elements of your life starting with "Isn't it amazing that..."

Our goal should be to live life in radical amazement... get up in the morning and look at the world in a way that takes nothing for granted. Everything is phenomenal; everything is incredible; never treat life casually. To be spiritual is to be amazed. – Rabbi Abraham Joshua Heschel

Monica Parker's work is part of the pantheon of researchers who have convinced me that being a Purveyor of Awe is good for us and good for the world (plus, she came up with that awesome word "wonderbringer"). She focuses on "wonder" rather than happiness as the antidote for the stressors and dissatisfaction of life. She believes that depending on happiness carries with it the danger of toxic positivity that narrows our range of emotions in unhealthy ways. Wonder, like my experience in the Snow Cat, can include a range of emotions from terrified to ecstatic. She has created a list of alliterations, five interlinking progressions for Wonder: watch, wander, whittle, wow, and whoa. Awe is the benefit of moving through these progressions, finally getting us to the states of "wow" and "whoa."

She describes "watch" as an "openness to experience." Being open and present means that we are "examining the familiar with new eyes to find undiscovered details and delights." People who nurture this quality consistently are found to be people who are naturally energetic, inventive, and compassionate. And guess what? Parker reports that people high in openness are more "wonderprone." Active imaginations, daydreaming, mind-wandering are all assets in this element. Spending time watching and then wondering internally means that we seek less external stimuli to fill the void. We listen to our own musings and are more in touch with our own ideas and feelings and sensations–all ingredients of wonder. I have a morning practice I call "CCC" (Coffee, Candle, and Contemplation) and it helps me with something Parker invites us to take seriously: "Open observation of our internal consciousness paves the way for curious exploration of our external world."

We live in a world where external stimuli lives in our pockets and buzzes and vibrates, demanding to be tended. If we are to be Purveyors of Awe who have identified wonder as an important element for curating spiritual depth, we must find ways to embrace and nurture this openness to experience, practicing unplugged daydreaming and mind-wandering and regular times of inward reflection. Our smart phones offer us a world of information whenever we want it. I am someone who loves research and is infinitely interested in how things work, what new things we can know about the world. I love being able to go searching any time a question crosses my mind. This is not antithetical to experiencing wonder–the world is amazing and seeking knowledge is not evil. Research can feed wonder-full ideas and thoughts into our lives. But as most of us know, there is a price we pay for this unending information. It threatens to demolish our ability to "exform."

My friends Cynthia Winton-Henry and Phil Porter developed a philosophy and technique of "Interplay" as a way of experiencing, knowing, and reflecting upon our lives and the world. I had the deep pleasure of playing with them for many years as a member of their Wing It! Performance Ensemble when I was living in the San Francisco Bay Area. This idea of "exformation" is seminal to their work. Through improvisational movement, making music, storytelling, and silence, we let go of the glut of information that is residing in our body-psyche-spirits so we can make room to notice what we ourselves know and think. It is what they call "body wisdom" and I have benefited greatly from this practice that helped me discover what was already clear wisdom for next steps in my life but had been clouded by external "voices" about what was right or expected of me. The paralysis of decision-making that I experience sometimes was able to let go into the wonder of what I already knew but had not allowed space to connect with.

> Without regular practices of "exformation" of all the "information" that comes in, we often don't leave room for "wondering."

Monica Parker's admonition to "watch" and to be open to new experiences requires some space for this "wondering." Being uninterrupted for a time is good for our mental states, lowering the fight-or-flight stress responses that come with "dinging" reminders that we've got a voicemail or that someone "likes" us. Our dog, Juno, can be woken abruptly and sent into anxious pacing with the ding of our phones or computers. So we keep our devices silenced all the time. It is a good reminder that this kind of reaction is also happening to us humans as we get neurological "spurts" of stress hormones every time we get a notification. Information addiction and existential dread ensue as regular physiological responses leave little room for the kind of "watching" with openness and anticipation that is the first step toward a life full of wonder. As Purveyors of Awe, we can invite ourselves and our communities to regular times of unplugging as a way to heal and offer healing to a wired-up world in need of awe-filled wonder.

Awe Practice: Wonderbringers

Dacher Keltner writes in his book, *Awe*, that in order to deal with the grief and "awelessness" he felt after the death of his brother, he "went in search of awe." He says, "I immersed myself as a newcomer in various wonders of life." As he did so, he became convicted that "awe is almost always nearby, and is a pathway to healing and growing in the face of the losses and traumas that are part of life."

The awe practice on this page is about naming, claiming, and experiencing a wonderbringer that is "nearby." In other words, what can you experience as if a newcomer that can pique your senses and elevate you for just a bit of time from things that are perhaps distressing or grieving you at this time. Think about your own multiple intelligences. Will it be related to words, music, visuals, spaces, movement? My mountains are no longer nearby so I went back to videos I made while skiing to reclaim that kinesthetic rush that I just don't get in the Midwest. Even that brought me a piece of that wonder.

Gently peeking its head around mental corners or bombastically announcing its arrival into our trembling psyches, wonder changes our perspective, our bodies, our souls, and our lives. Art, music, religion, politics, science, nature, love, fear, birth, death; each of the myriad experiences that compress to form the bedrock of human life has a golden vein of wonder running through it. – Monica C. Parker, The Power of Wonder: The Extraordinary Emotion That Will Change the Way You Live, Learn, and Lead

Meaning

We humans are "meaning-mongers." Our brains are wired to interpret the ordinary stuff of life in metaphorical ways so that we can make some sense of it all. We long for meaningful lives in which we are more sure about the direction of our lives and about what we are to bring to this world. We yearn for experiences that leave us feeling more fulfilled. When we are on the lookout for meaning in our lives, we live at a deeper level and mere existence turns into purpose.

I love New Year's Eve. It is such a liminal, "betwixt and between," moment. I have had lots of New Years Eve parties over the years made of meaningful rituals that people could do to contemplate the passage of time more deeply than just the usual mix of alcohol and kissing at midnight. This year I offered a New Year's Eve ritual for the neighborhood–a kind of come-and-go outdoor experience. I call these kinds of easy-access rituals "casual encounters of the spiritual kind." The sign I put next to a bowl of beach glass said this:

"Take one or more pieces of beach glass from this bowl representing what you do not want to take with you into 2025. Walk up the pathway to the 'circle of months.' Contemplate the year 2024. Put the glass into the bowl at the center as a sign of release. Walk away into a renewed start to 2025!"

I communicated to people through our neighborhood social media that this was available. I also watched as people happened upon it all through the night as they walked past the house. All expressed surprise and gratitude at being given the opportunity for a moment of meaning.

The language of symbol and ritual is a doorway to meaning. We can only know mystery through the tangible stuff of our lives. And we can only appreciate the act of "knowing" when we reflectively experience with awe that we are beings who can actually know anything. That's what Lily Tomlin's "Trudy the bag lady" character in Jane Wagner's play "The Search for Signs of Intelligent Life in the Universe" tells us about awe and understanding awe:

"And then I felt even deeper in awe at this capacity we have to be in awe about something. And I became even more awe-struck at the thought I was in some small way a part of that which I was in awe about. And this feeling went on and on and on and on. My space chums got a word for it, 'awe infinitum.' 'Cause at the moment you are most in awe of all you don't understand, you're closer to understanding it all than at any other time. And I felt so good inside, my heart felt so full, I decided to set time aside each day to do 'awe-robics.'"

"Awe-robics" is what we need for a meaningful life. especially when we let go into the wonder of what it means that we are capable of feeling this "awe infinitum." Like Trudy, we have to go all-in.

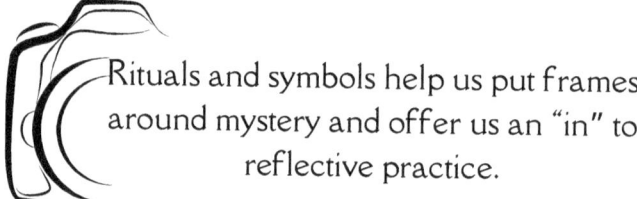
Rituals and symbols help us put frames around mystery and offer us an "in" to reflective practice.

Ritual and symbol are based on metaphors. We experience the tangible but then our brains connect it in a much larger way, seeing a life lesson evident in an ordinary thing. I like to forage for metaphors, thus, many years ago I came up with the term "metaphorager." As Purveyors of Awe, I want to invite you to be a metaphorager as well. Let me give you an example from my life that demonstrates how metaphors deepen meaning: my first tattoo.

When I was about to turn forty years old, I wanted to do something I thought I would never do... and I was not about to jump out of an airplane! So the other thing that, as a younger person, I had thought I would never do was get a tattoo. I mean, how can you decide something that will be on your body forever?!! But then as I began to take in the reality of getting older, I realized that "forever" is actually not a very long time! My body was beginning to be marked with age and so I figured I would start marking it up myself as well. My father had recently married a woman from Guam, born in Taiwan. I asked her to write the Chinese letters for "peace and passion" (the words I used in benedictions) and I had them tattooed on my ankle. She gasped when she saw it the first time and I wondered if I should have checked it out with another Chinese person before getting the tattoo. Did it really say "stupid step-daughter?" No, she had gasped because it was her actual handwriting, this was an exclamation of awe. Now that she has passed on from this life, the tattoo is a way she is still present as well as a symbol of my birthday rite of passage and my connection to my vocation through the choice of words. Symbols have the ability to make things more present to us and create layers upon layers of meaning.

This is the purview, as author Brian McLaren calls it in his book, *Life After Doom: Wisdom and Courage for a World Falling Apart*, of the "meaning committee" inside our minds. He describes three committees that follow the evolution of our brains: the *survival committee* is the most primal development whose purpose is simply to keep us alive; the *belonging committee* developed as humanoids needed strong bonds with each other for survival; and the *meaning committee* "gives meaning to the word 'meaning.'" It is a clever short-cut description of a much more complex story of human evolution, but helps us "map our minds," McLaren says, so that we can "guide and curate and manage it."

As we think about curating a life of spiritual depth, this model can help us reflect on the dance of these three committees in our lives. Our very survival may not be at stake and yet if we set such high standards of what "survival" is, we may work ourselves "to death" and right out of time for nurturing a sense of belonging and meaning. If we are so bent on "belonging" that we develop animosity toward those different from us, we may undermine greater flourishing for all people that comes with a sense of the greater whole, thereby endangering the survival of our species and the planet. Spending time curating meaning can help us put survival and belonging in perspective if we don't simply relegate meaning to a navel-gazing endeavor. Each of these things–survival, belonging, and meaning–are important to nurture. Perhaps this is what Brother David Steindl-Rast means when he says "It is through wholehearted living that meaning flows into our lives."

Is there a story of something that serves as a symbol of a deeper meaning for you? What is the role of meaning-making in your life? What is the conversation between the survival, belonging, and meaning committees in your mind?

A lifetime may not be long enough to attune ourselves fully to the harmony of the universe. But just to become aware that we can resonate with it–that alone can be like waking up from a dream. — David Steindl-Rast

The meaning-making ability that our brains have developed can also become tedious and tyrannical without the help of the release into awe. My first adult career was in professional dance, as I've mentioned. But very specifically, I was in two major professional abstract modern dance companies. While some pieces we performed had the hint of a story line or characters, most of the pieces were simply explorations in motion with particular stage, costume, music, and light settings that would set an ethos (a kind of world) and a "feel." This sensory-rich combination would act as the "essence of an experience" for the audience rather than tell a story such as a narrative ballet or opera or musical theater piece does. Both artistic expressions–the abstract and the narrative–are equally powerful but different. I think that abstract artists offer us a mirror to see how most of our sensate life really is. We glean meaning not always by our particular story lines but by moments that simply add value to what it means to be sentient beings.

> A melody line conjures a feeling. The angle of the light conjures a memory we can't quite put a finger on.

When someone would ask "what did it mean?" in reference to a particular modern dance piece, we would ask right back, "what did it mean *for you*?" What anyone in the audience interpreted it to mean was just as valid as what it meant to the artist in its creation. Often an abstract artist is playing with line and color and harmony and shape and form so that they can offer something for others to play with, creating their own meaning as the piece crosses paths with the onlooker's life and experience. That is the point–the interaction and the improvisational meaning-making of the witnesses. What if the Creator did just this?

> **The heavens herald your glory, O God,**
> **and the skies display your handiwork.**
> **Day after day they tell their story,**
> **and night after night they reveal the depth of their understanding.**
> **Without speech, without words, without even an audible voice,**
> **their cry echoes through all the world,**
> **and their message reaches the ends of the earth...**
> **Holding you in awe, YHWH, is purifying; it endures.**
> –Psalm 19 (Inclusive Bible)

Perhaps faith is not about landing on the right dogma about "the way things are" but rather invites us to the improvisational dance of awe that produces the feeling of being deeply a part of this creation. This, then, is the depth of meaning. It produces not grand linguistic theses but a profound and enduring knowing beyond words. Too often we seek to "get it right" when perhaps it is about simply "getting it," whatever the "it" is for us within our current framework of knowing. And when that framework is altered by life experience and exposure to other insights, the "it" we get changes. Overemphasis on right answers and knowledge as data-driven has created a sense that our meaning-making must produce something solid and eternal rather than offer us the opportunity for the unending improvisation that is the *mysterium tremendum* of awe at simply being alive.

In my dance career I also spent five years with a dance company that toured with the Dave Brubeck Jazz Quartet. We played to sold-out theaters around the world, filling them with both dance and jazz lovers, sparking discovery for each of them with the combination of both of them. There was an energetic explosion at the end of every performance as we dancers rushed the edge of the stage with the final notes of "Take Five." The meaning was the magic when people came away in a kinesthetic high that defied description–except perhaps "that was so awesome." And I can tell you that every night was just as meaningfully magical for us, the performers, as it was for those who had witnessed it from their seats.

The show was filled with set choreography to specifically-arranged versions of the Brubeck's greatest hits. But then we would let loose at the end of "Take Five" with improvisational solos–musicians and dancers. Even better were the master classes we co-taught with the Brubeck Quartet in music and dance departments of universities across the country. It was in these sessions that we really had time to relish the art of creating in-the-moment. Improvisations through performance art have been, for me, some of the only times when I truly release what is a kind of "tyranny of meaning-making" when angst about the past or worry about the future colonizes my attention and over-active imagination. Improvisation demands an all-in of being present to the moment.

Making meaning isn't about certainty, it is about play. If I think of life as an improvisation then, like Trudy, "awe-robics" will keep me playing, will keep me present, will keep me whole.

Awe Practice: The Camera Lens

The camera lens I've been using as a graphic in this study is itself a metaphor. We are looking through the lens of awe to see the world and others; looking deeply to receive anew with wide-eyed wonder the precious nature of the world and our existence in it. When we create a photograph, we spend time "framing" the shot–we focus in on it in order to represent its essence and communicate a particular perspective. We are refining our gaze to glean a sense of meaning in it.

In this awe exercise, I invite you to be a "metaphorager" by utilizing an actual camera. Don't wait for something extraordinary to cross your path because there are ordinary things around us constantly that can reveal awe if we only tune into them. Just keep your spirit attuned and your camera at hand. When something "captures" your eye, frame it, zoom in on it, find the essence of what drew you to it. Snap a shot, or a few. Don't walk away before noticing how you feel. Then spend time later looking at your photos, imagining and playing with what meaning may be presenting itself.

Without the meaning committee [in my brain], there would be no Bach or the Beatles, no Galileo or Einstein, nor Moses or Mary or Jesus or the Buddha or Mohammed. It is the most recently evolved part of me, and frankly, it still has a lot of bugs and glitches that haven't been worked out yet. – Brian McLaren, Life After Doom: Wisdom and Courage for a World Falling Apart

Curiosity

As I write this we are moving through a time of disturbing foundation shaking in the United States in which structures that uplift the most vulnerable are being broken down. Each time I return to my writing I ask myself, "does focusing on awe really matter?" And then I begin to review the research and I am reminded that these attributes of curating a life of spiritual depth are not "fluff." They are about our very ability to exist and thrive, to be in touch with what we value in order to discern next steps, to be equipped for the healing that must happen in order to answer the challenges we are facing. The element in this section is curiosity–not simply the information-seeking kind, but what researcher Scott Shigeoka, author of *Seek*, calls "deep curiosity." After years of studying curiosity, he believes it can become a force for meaningful connection and transformation.

Curiosity has been a part of survival, healing, and transformation for me personally. A few years ago, I set one notification on my phone that popped up at 10 a.m. every day for over a year. It simply said "Wonder."

The notification acted as a disrupter for my brain, which, at that time was obsessively worried and anxious about the wellbeing of a deeply troubled relationship. The word "wonder" was a double entendre. It served to prompt two things: the first was to get out of my head and look around to see what wonders may be right in front of me in order to alter my mood. The other purpose was to incite curiosity–it was a prompt to stop and identify the worry I was spinning out on at the moment and turn doom predicting into a question, *"Huh... I wonder how that will turn out?"* It was a way to transform debilitating anxiety into curiosity, and it was a sanity-saver for me.

I am a big fan of having strategies–practices–to help us exercise the muscles that will invite wellbeing. The muscle of deep curiosity is one of the most important to exercise, in my opinion, based on experiences like the wonder reminder on my phone. It opens up possibilities for more awe in our lives as we learn to meet uncertainty with curiosity. Scott Shigeoka took a trip across the United States as a way of practicing the tenets of his research on how to bridge the differences in a world increasingly full of animosity and, as he puts it, an "incuriousity" that is killing us. What he found is that curiosity has many benefits for individuals and society:

- Challenge our assumptions and biases
- Provide an antidote to fear and anxiety
- Embrace uncertainty with more courage
- Deepen connections in an era of social isolation and exclusion
- Become more intentional and thoughtful
- Sharpen our creativity and collaboration skills
- Find common ground with others who have opposing views or differences
- Move through hard times in our life
- Build self-awareness and be kinder to ourselves

See what I mean? Every time I wonder (ha!) if these elements are worth talking about in the midst of what feels like global existential crisis, I run across someone who has been doing work to prepare us for "such a time as this." Each of the benefits above are collectively necessary to survival: emotionally, spiritually, physically.

> Curiosity requires a humility that recognizes we don't really know unless we ask. And a courage to actually ask.

Shigeoka reminds us we need to be "seekers." Strategies for this practice include the honing of four actions (DIVE): **D**etach from our assumptions, biases, and certainty; **I**ntend by preparing our mindset and setting; **V**alue the dignity of every person, including yourself (which means there are instances where it is necessary to protect rather than extend ourselves); and **E**mbrace the hard times in our lives. This is no small task, and we are feeling that in no small way in this moment. The creativity needed to find solutions depends on a curiosity that requires humility and courage on the part of everyone.

Meister Eckhart is my favorite late 13th century theologian and mystic (well I don't know that many, but if I did, he would be my favorite I'm sure). A quote I have used often is this, "Every creature is a Word of God and a book about God." The translation, using liturgical terms, is this (see me looking into your eyes): "You are a Word of God. Thanks be to God." There is an essence in each person, says the mystic, that can reflect an understanding about the intention and presence of Divine Love in the world. It may be buried deep beneath layers that seek to keep it strangled and silent, but it is there. The only way we get to unearth it in ourselves and in our perception of others, is deep curiosity. "Be willing to be a beginner every single morning," he said. Eckhart, who was accused of heresy, wrote in a time of great turmoil and inquisition within the church to ferret out any "irregular" beliefs. He spent time with lay persons and translated his writing for them into their own vernacular language from the prescribed Latin of the church. Understanding was a holy endeavor for all, not just for some.

The Hebrew poet agrees. We are to dig for wisdom like a buried treasure:

> **Incline your ear to Wisdom,**
> **and take her truth into your heart.**
> **For if you yearn for insight**
> **and cry out for understanding,**
> **if you search for it as you would for silver,**
> **and dig for it as you would for buried treasure,**
> **you will understand what awe of YHWH is,**
> **and discover how to truly know God.**
> – Proverbs 2 (Inclusive Bible)

We cannot give up the search. Being a true seeker takes three directions, says Shigeoka: inner curiosity (what makes me tick), outer curiosity (what makes others tick), and a curiosity toward "the beyond" (connection with something bigger than just us). "Explorations of the beyond can help you to feel more grounded in the here and now," he says. We might be more comfortable with one of these three directions than the others, but cultivating deep curiosity in all three directions is essential. Inner curiosity helps us avoid a lack of self-awareness, outer curiosity wards off narcissistic tendencies, and curiosity toward the beyond can negate despair that comes when we lose touch with the bigger picture.

What can you get deeply curious about that would help transform a fear into an inquiry, a dread into an investigation, or a bias into an exploration? How does curiosity show up in your life: inner, outer, and into the mystery of the beyond?

Instead of operating from a place of fear, trauma, or scarcity, we can unlock a sense of security, joy, acceptance, ease, play, awe, courage, intimacy, and freedom. That's why I call this version "deep curiosity." You dive beneath the surface. – Scott Shigeoka, Seek: How Curiosity Can Transform Your Life and Change the World

In an age of "knowing" exemplified by the kind of standardized testing that infiltrated my early education, curiosity, creativity, and critical thinking have taken a hit. I've always had an off-the-charts imagination (good for creativity, bad for anxiety) and am hounded by a muse who gives me multiple big ideas every day. But I sucked at standardized tests. Perfectionism and performance anxiety meant that school, which I usually adored, would become terrifying during test time. Over the decades since its inception, the "wisdom" of such a focus in education has come under question. I think we may be seeing the fruits of generations of people who always look outside themselves for the right answer–taking whatever is presented at face value–rather than relying on developed curiosity that invites them to weigh the evidence and imagine a larger picture than the one which they are presented. Now that we are mixing AI into the equation, the question of the relationship between data consumption and human creativity is ever more important.

Deborah Farmer Kris is the author of a new book, *Raising Awe-Seekers: How the Science of Wonder Helps our Kids Thrive*. She believes that rediscovering awe can boost kids' mental and emotional well-being, strengthen their social ties, supporting their curiosity and internal motivation. Drawing on Dacher Keltner's *Awe* research, she makes suggestions for parents and teachers who help children face the intense challenges of today's world: slow down childhood; embrace playtime, downtime, and family time; practice radical curiosity; and become an awe-seeker yourself.

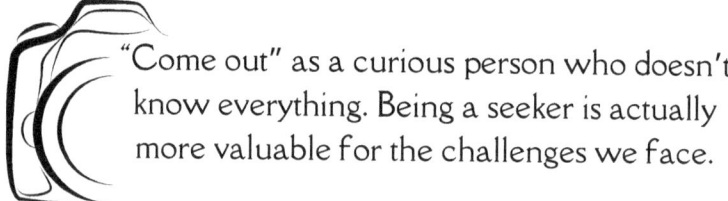

"Come out" as a curious person who doesn't know everything. Being a seeker is actually more valuable for the challenges we face.

Shigeoka agrees adamantly with Farmer Kris' last suggestion: become an awe-seeker yourself. In his last chapter of *Seek*, he makes suggestions for creating a more curious society. And it starts with each of us "coming out" as a curious person. The closet most of us are in is the one where we pretend to have the answers. We do this because our society has placed so much value on *knowing*, attaching promotions, scholarships, and labels of "smart" upon those who do well with standard measurements of intelligence. But the genius Albert Einstein famously said, "The true sign of intelligence is not knowledge but imagination." And imagination thrives on curiosity.

If we wonder how to invite more curiosity in those around us, the good new is: curiosity is contagious. The more we practice deep curiosity and not just consumption of information at face value, the more those around us will tend to do so. The "mirror neuron" phenomenon studied by neuroscientists and psychologists means that we have the capability to invite others to shifts in attitudes and practice not by convincing them through grand and persuasive statement or argument, but by simply *being curious* around them. This neural mimicry lights up the same areas of their brains just by observing our deep curiosity practices.

In Monica C. Parker's work on *The Power of Wonder*, her second "W" is "wander." She is speaking of what she calls "wondrous curiosity." It is different than the surface curiosity that evokes bursts of "Google-It!" Wondrous curiosity contains a love for the exploration, an enjoyment of complex and nuanced answers, and has a meaning-making purpose. It is learning as pleasure and induces the same kind of dopamine reactions in the same parts of the brain as romantic love. Wow. That's quite an addition to the benefits of curiosity. And yet, like anything that triggers dopamine release, we have to be mindful that the same tendencies can also create an insatiable appetite for more, such as more doomscrolling that is specifically engineered to target our body's craving for a dopamine high.

Both mirror neurons and dopamine highs are examples of just some of the awesome abilities of our brains that can work for us or against us. Our innate tendencies can go unexamined and unchecked leading to unhealthy habits. Or we can "curate" them. We can choose those things that we know help us and the world around us. And we can invite our families, friends, neighborhoods, and larger communities to engage in practices that create the bedrock of spiritual depth to propel us into a better future.

This day of writing comes to a close and I reflect on my opening statement after having read this morning about some pretty disturbing world affairs. I feel more resolved about whether or not focusing on the elements of awe such as curiosity matters in times like these. I remember one of Brian McLaren's points in his list of "what matters and what matters even more" at the end of *Life After Doom*: "What you have already learned matters, and remaining curious matters even more." So perhaps it is time to start the phone notification practice again. This time it says "Stay curious."

Awe Practice: Curious Conversations

I love to travel. One of the things I love about it is meeting new people because every encounter is an opportunity to learn a different perspective and to be changed by the encounter. Travel expert Rick Steves says that strangers are simply friends we have not yet met. Yes! I am in awe, to use a phrase, of the deep sharing I've had in encounters with so-called "strangers." Whether on a bar stool in Ireland or a seat on a plane, there are conversations that have "blown my mind" with their poignancy.

We can't really predict when a conversation will dive into the deeps but we can put conditions into place that support deeper sharing. Offering our own vulnerability as well as our deep curiosity and openness to listen and learn are the practices for this section. Instead of asking "what do you do?" ask an "into the deep" question such as "what gives you joy these days?" Just see what happens.

The search is what anyone would undertake if he were not sunk in the everydayness of his own life. To become aware of the possibility of the search is to be onto something. Not to be onto something is to be in despair... I have discovered that most people have no one to talk to, no one, that is, who really wants to listen. When it does at last dawn on a man that you really want to hear about his business, the look that comes over his face is something to see. – Walker Percy, The Moviegoer

Delight

We all have places where we sense a closeness to the Divine–however you describe that. We get a "feeling" from these places–a sensation, a very real physical phenomenon like goosebumps or breath "taken away" or a deep longing filled. I have a phrase when I have such experiences: my "delight bucket" is filled up. In the first few months of the pandemic lock-down, I realized that I was getting depressed. I'm not prone to prolonged depression but "Lord knows" there was enough to be depressed about. When I realized that I needed my "delight bucket" refilled, I knew I was onto something. And sure enough, curating some experiences that I know are delightful for me was the antidote for what was ailing me. Paying attention to the elements of awe by curating moments that place us in the path of delight, is important to our wellbeing and, therefore, consequential for the wellbeing of those around us. Philosopher and filmmaker, Jason Silva, invites us to think of our lives as a "work of art" and ourselves as the curators of the spaces into which our lives "unspoool" as a film reel does. He invites us to "pre-design" experience by placing ourselves in contexts of delight.

I love curating light. I think fluorescents are "of the devil" (I don't actually believe in the devil but if I did, hell would be lit by fluorescent lights). Obviously, my sensibilities run toward the warmth of candlelight, lamplight, and the magic of theatrical lighting. One of the companies I danced with was world-renowned for its cutting edge light techniques. We toured with dozens of slide projectors (I know, that's how old I am) that flooded the stage and us dancers with amazing color and shape, thrilling audiences with mind-blowing environments onstage that they had never seen before. I remember experiencing the company perform when I was in college and being so wowed in my seat in the balcony that I vowed I would place myself on a trajectory to be a part of that company. It was pure delight for me and I didn't want to just watch it, I wanted to be *in it*. Ever since, I've been playing with light inside and outside buildings, on stage and in churches–now using computers and LED projections (much easier than slides).

And so during the pandemic, I curated lighted spaces indoors in our home that would be full of *hygge* (more about that in a moment) and visited magically lit outdoor places such as botanical garden light shows and holiday drive-throughs, knowing the effect of transcendence that it created within me which helped counteract the feeling of pandemic "lockdown."

The Hebrew poet helps us express our awe at the delights of creation:

> **How precious is your love!**
> **Whether creatures of heaven or children of earth,**
> **we all find refuge in the shadow of your wings.**
> **We feast on the bounty of your estate,**
> **and drink from the stream of your delights.**
> **In you is the wellspring of Life,**
> **and in your light we become enlightened.**
> – Psalm 36 (Inclusive Bible)

Whether it is a well-lit room, a particular place in nature, being around a table of delicious food with friends, a puppy-pile with your kids, a bouquet of flowers, or a gazing up at a starry night, we all have things that feel like a "wellspring of life" and we are invited to drink deeply from the stream.

I had good teachers for this in my childhood. My parents curated our lives with intention. Even though my father was a pilot for a commercial airline, we moved to a farm in Missouri because my parents wanted my brother and I to have the experience, as they did, of growing up in a small town with the joy and responsibility that raising animals and growing crops offers. And they made sure to sprinkle our lives with delightful activities from driving to "the city" for Royal's baseball games, magical brush fires in the pasture, teaching one-day-old baby chicks to drink, collecting honey from beehives, and riding rodeo on my horses. And of course, my mom driving me an hour each way to dance class, without which my love and skill would not have ended up in a professional career.

So as they both age, I have wanted to give back moments of delight to them. My mom was with us for a time during the pandemic and one of our excursions was to a botanical gardens that had installed a light show timed to music (oo-la-la). I have a video of one of the installations. I didn't realize until I watched it later on a bigger screen, that the video included a shot of my mother watching with the wide-eyes of wonder, somewhat childlike and a reminder of how wonderfully unabashed she is at moments of delight. I'm so grateful to have had that kind of "be astonished" role model and a video clip of her which I will treasure always.

I recently took my dad to Ireland. While he has logged thousands of hours seeing earth from 30,000 feet up, it is the earth itself as he studied it in an agriculture degree in college that enraptured him during this visit to a land where our family first learned its farming techniques. I love his frequent exclamations of "That's amazing!" about things from a hand in a card game to watching the prowess of a sheep herding dog, clearly his favorite part of our trip.

> A penchant for awe, wonder, and delight are as much an inheritance that we can hand on as any amount of money.

Curating the conditions for delight has familial and communal implications. One of my favorite films is *Babette's Feast* in which the main character, a French woman, is taken in during wartime by a small Danish community. The people of the village are very strict and deem any pleasure a sin. They wear bland clothes, they eat bland food. They are in a long-standing family feud. There is a dinner hosted by the French refugee in gratitude to the community for the gift of a safe place to live during the war. What the town finds out is that this woman is a chef and the gathering will include a sumptuous multi-course meal. They want to honor her gesture so they decide they will go, they will eat, but they will not enjoy it! The scene of the meal in the film is a slow roll of transformation. They cannot help themselves as they slowly give way to the delight of the meal. The act of sharing even sparks a reconciliation between feuding families. Indeed, the act of setting a table, preparing a meal, exercising hospitality is itself a symbol of the power of curating spaces where transformation can happen. When our "delight buckets" are filled, this sense of abundance creates a generosity of spirit.

Who taught you to delight in the world? What are your earliest memories of "drinking from streams of delight?" How could you curate contexts for the "unspooling" of your life experiences now?

Our tears register our awareness of vast things that unite us with others. Our goose bumps accompany notions of joining with others and facing mysteries and unknowns together. Today we may sense these laws of bodily awe when moved by a favorite musical group, or in calling out in protest with others in the streets, or in bowing our heads together in contemplation. And in such rushes of tears and chills... we may glean a sense of what our souls might be. – Dacher Keltner, Awe: The New Science of Everyday Wonder...

The work of the soul is also the work of the body. The fact that many of our moments of awe, wonder, and delight are accompanied by physical reactions such as goose bumps, hair follicles standing on end, and sometimes even getting "teared up" affirms the very bodiliness of spirituality. Your mind is not just the conscious analytical mind. Nor is it just located in your head. You have a mind in your body. Your body is a mind. It is an incredibly complex set of feedback systems working in concert with brain functions. Equally true is our experience of the "soul" or "spirit." We are one entity because there is no way to sense any experience–physical, mental or spiritual–without our bodies. And so the experiences of our bodies will have an affect on what we think and feel and how we understand the "sacred" or the "soul." What constitutes the feeling of oneness, of deep joy, or of profound knowing is intimately tied to the experience of our bodies.

One of the most delightful experiences for me that comes out of growing up on a farm is connection with animals. I used to hang out in the barn, especially on winter mornings after breaking the ice in the water trough and feeding the cows and horses their hay. The physical feeling I remember is a slowing of my heart rate, a deepening of my breathing, and a sense of calm–no small thing for a generally anxious and overachieving kid. That memory popped up for me recently when I visited my local donkey rescue farm. I wrote an Advent/Christmas worship series that featured the character of the donkey and so, as usual, I went to my research happy-place (my spouse and I have alter-egos named "Beulah and Daisy" because we are happy as pigs in s#!t when researching something) and did a little digging about donkeys, including this visit to experience them first hand. And let's just say that now I am obsessed. The rescued donkeys at Zen Donkey Farms have a special role in hanging out with kids from the local children's hospital. Research shows that donkeys can help with everything from calming nerves to helping children with autitism learn to read social cues. At the donkey farm, I experienced sighs of release, tears of joy, and the medicine of laughter at the hilarity of their play.

When our humanoid ancestors experienced unexplained phenomenon, vastness, and perilous dangers, their goose bumps served to draw them closer together, offering warmth in proximity and connection via the touch of huddling together. Even though we modern humans don't face down tigers on a regular basis and we actually know what is happening when thunder clashes, we still can draw on this tool for the more prevalent danger to us in this era–the threat to our mental health. Studies have shown that stimulation of the vagus nerve, which Keltner describes as a key player in sensations of awe, has implications for the treatment of depression and PTSD, among other mental ailments. The rescued donkeys, many of whom healed from their own trauma by being with other rescued donkeys in a nurturing environment, now pass that proximity and sense of connection on to the children and adults whose nervous systems need a good dose of delight.

> Creating a space in which to reset and reconnect with delight may come down to a nook, a book, a candle and a plant.

The Danish have a word that has defined their lifestyle and spread recently out into the world: *hygge* (hoo-ga). Meik Wiking is a researcher based in Denmark who studies the "happiest places on earth," which happen to be some mighty cold and dark northern European countries, including his own, especially in the winter. Perhaps their contentment comes out of the need to curate good vibes to survive those long months. Thriving in those conditions can serve as inspiration to pay more attention to curating our own spaces and analyzing how we feel in them, what "vibe" might be missing for us, and doing something about it. To quote Wiking from his *Little Book of Hygge*, we need to get "consciously cozy."

Defining *hygge* is not easy as it, like awe, is about a feeling. But the Danes have some things that, interestingly, have shown up consistently over time and are very concrete and repeatable. The word for "candles" in Danish is translated as "living lights" and eighty-five percent of Danes say candles are the number one necessary ingredient for a sense of *hygge*. Ah... my people. And they are adamant about choosing lamps for their homes that create soothing pools of light (I'm guessing fluorescents are not big in Denmark, another mark in the plus column). Curating soft textures in cozy spaces, making a cup of tea and a sweet in the afternoon, valuing natural materials in the home such as wood and plants, are all very *hyggelige*. But the goals for creating such delight are not just about personal coziness. Here are ten aspects of the Hygge Manifesto: atmosphere; presence; pleasure; equality; gratitude; harmony; comfort; truce; togetherness; shelter. The word *hygge* comes from the Norwegian word for wellbeing. Danes pay some of the highest taxes (they call it "investing in quality of life") and have the most robust welfare model. The right to delight is for everyone.

Awe Practice: Delightful Spaces

Inspired by the Danes, this awe practice has to do with noticing, evaluating, and sprucing up our living and working spaces for optimum *hygge*. Think about the rooms or nooks in your home or at work that you spend the most time in. Do they that give you a sense of vitality or offer a bit of respite? What would improve them? For instance, I started using small lamps at my desk instead of overhead lighting because it helps me feel held when I'm needing to deal with the anxiety of finances or deadlines.

Then think about creating some *hyggelige* moments of community. If cooking is your jam, when is the last time you cooked for an intimate gathering of friends? Or is there a restaurant that is particularly cozy where you could "hole up" with friends or family? How would you make the conversation more *hyggelige?* Perhpas creating boundaries about current events and instead having a round of recalling a moment of delight in your lives could be a good start.

We are slowed down sound and light waves, a walking bundle of frequencies tuned into the cosmos. We are souls dressed up in sacred biochemical garments and our bodies are the instruments through which our souls play their music.
– Albert Einstein

Connection

I had a huge "aha" many years ago at the first mega-event I got to help design and curate–a gathering of 12,000 United Methodist Women. We were in an indoor stadium setting. One evening was slated for a concert with one of my all-time favorite music groups–the Paul Winter Consort. Their music is on my awe-inspiring music playlist with their merging of recordings of animals in the wild with Paul's soprano saxophone and the other incredible instrumentalists in the group. I was already primed for an evening of awe, wonder, delight, meaning, and the rest. But the "aha" moment came as an unexpected experience of connection.

The performance was delayed about 15 minutes. So the stadium was full and waiting. The vibe was abuzz with people talking to their neighbors, pretty much focused on the person right next to them or in small groupings of adjacent seats. But then a small section began to try starting a "wave." You know, the kind you see at sporting events when each section undulates their bodies up and down as the wave reaches them.

It took a few minutes to catch on as people began shifting their attention from personal conversations to what was transpiring in the room. But then it happened. The wave began and 12,000 people created a surge of connection that lit the place up. In the space of a few minutes we went from being thousands of individuals in the same space to making a connection that created a sense of the whole as one body. It was organically orchestrated magic and resulted in communal "whoops and hollers" (as we say in the Midwest). By the time the Paul Winter Consort took the stage, we were no longer sitting back in our seats, we were energetically leaning forward.

As social animals, we are especially wired for awe when it comes to communal connection. My first story of walking to the beat of music in New York City resulted in a unitive state–a suspension of my sense of the boundaries of self and a greater connection with "all-that-is." This is possible on one's own but it is multiplied many times when we experience it with others. We can feel a sense of merging with others resulting in what Keltner calls an "ocean of awe."

This was evolutionarily important as we humans began to bond in groups for greater chances of survival. In William McNeill's study called *Keeping Together in Time*, he explores the ancient phenomenon of "muscular bonding." He believes that in the course of human history, *homo sapiens* learned to physically "keep time together," which gave them a great evolutionary advantage. In their ritual dances around the fire, their common rhythm and motion helped them feel more power, contributed to their health and offered preparation and energy for the hunting season. In their walking from place to place with songs and rhythmic stepping, they felt a sense of belonging to a whole along their journeys. With their songs as they were ground corn and wove fabric, they could work longer as the rhythms carried them along.

I would say that as Purveyors of Awe, we must curate connection practices, both on a smaller scale to combat the epidemic of loneliness, but also inter-tribally in larger occasions of synchronicity to combat the illusion of our separateness from each other and from creation itself.

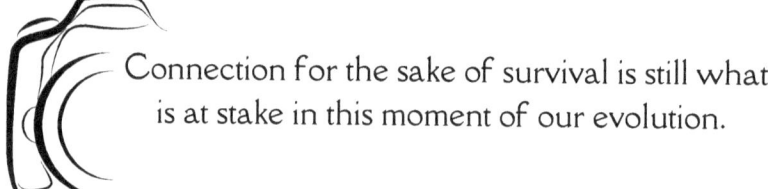

Connection for the sake of survival is still what is at stake in this moment of our evolution.

In my PhD work on the neurophysiology of ritual, my awe at the human body and its capacities escalated, especially as I wrote about energy dynamics of bodies in action (singing, dancing, walking, gesturing) together. The patterns of energy dynamics produced in these instances create a synchronization of limbic discharges (one of our most ancient capacities and the seat of emotions) between individuals. This can create states of being within the brain which blur the differentiation of self and other. These states of being created by the limbic system within our bodies are shared between individuals in a group body which experiences a "synching up"–a kind of kinesthetic identification.

This phenomenon was termed "collective effervescence" by social scientist Emile Durkheim in 1912 as he studied indigenous rituals. He observed that moving in unison elevated the participants to an exalted feeling and a sense of shared awareness of unity and meaning. He called this effervescence the "soul of religion." If we take Durkheim's definition to heart, and we think of the root of the word religion, which is to bind together, then our synching capabilities are the embodiment of our soul work together.

Modern day researcher and professor of psychology, Shira Gabriel, continues the investigation of collective effervescence in her study of the effects of large singing events for human flourishing. Events such as those hosted by a group from Israel called "Koolulam" where hundreds and sometimes thousands of people join in a mass effort of creating a sung piece together, have a strong correlation to less stress, lower depression, and a higher sense of the meaning and purpose in life, says Gabriel. Indeed Koolulam's stated mission is for "strengthening the fabric of society." Its very name is a combination of an Arabic word *kooloo*, meaning "ululation" (a kind of singing) and the Hebrew word *kulam*, meaning "everyone." Even the marriage of the two languages in their name is an indicator of their goals as a social musical initiative.

When instances of collective effervescence are layered with shared meaning and a shared cause, the result can have ripples that last for a long time in the people's lives who have experienced it. This is nothing new, of course, because this is the essence of ritual strategies of singing and moving together throughout human history. However, what Koolulam does in bringing together people of differing backgrounds, languages, and religions is a courageous, and I would say, "medicinal" act essential to our survival as a global community. They invite people out of their tribal silos for a time. When Jews, Christians, and Muslims sang Bob Marley's "One Love" in English, Hebrew, and Arabic in the shadow of David's Tower, the experience gave each of the 1000 people a sensate memory of a lifetime and gave the world an aspirational vision of who we are as a human family. As Purveyors of Awe inspired by poets, researchers, and practitioners, can we answer a commission to create and support efforts like these that over time might just make a difference for flourishing of all?

When have you felt a sense of awe at the connectedness of the human family? What impact did the isolation of pandemic have on you? What effect does political polarization have on your sense of wellbeing?

Experiences [of collective effervescence] contribute to a life filled with less loneliness and greater meaning, positive emotions, and social connection... Collective assembly meets the primal human yearnings for shared social experiences. A collective assembly can start to heal the wounds of a traumatized community. When we come together to share authentic joy, hope, and pain, we melt the pervasive cynicism that often cloaks our better human nature. – Brené Brown, "Why Experiencing Joy and Pain in a Group Is So Powerful"

The Hebrew poet underscores the power of acts of collective expression:

> **Alleluia!**
> **Sing to YHWH a new song!**
> **Sing praise in the assembly of the faithful.**
> **Let Israel be glad in its Maker;**
> **let the children of Zion rejoice in their God.**
> **Let them praise God's Name with festive dance;**
> **let them sing praise with timbrel and harp.**
> **For YHWH loves the people,**
> **and crowns the lowly with salvation.**
> – Psalm 149 (Inclusive Bible)

The effect of the assembly's connection is an experience of unitive love and a healing "salve" for the vulnerable and oppressed. We see in this psalm the ways that praising together might have provided the necessary balm for the pain of exile and joyful catharsis in celebrating their return. Layne Redmond, of blessed memory, and the author of *When the Drummers Were Women*, said "falling in love is falling in rhythm." When we experience the pain of disconnection, we fall out of the rhythms that felt familiar. We may feel listless and out of sync, "strangers in a strange land," until we get our groove back.

As a card-carrying codependent (and mighty grateful for the CODA 12-step program), I know the deep pain of disconnection when faced with the detangling of toxic relationships with people and institutions. Creating healthy boundaries and trusting connection again is difficult. And the pandemic created disconnection on a scale that I believe we have not yet begun to understand.

> Healing of our personal and collective trauma through healthy connection is part of the work to regain our capacity for awe.

So much is at stake. Purveyors of Awe are called to create connections whenever possible–large or small. Indeed, more intimate gatherings are where we are likely to have more frequent opportunities for the healing journey through togetherness. When my spouse and I bought our new-to-us hundred-year-old house, the first piece of furniture we bought was a long rosewood table that would fit a gaggle of people around it in our dining room. This was part of our commitment to each other for this next phase of our lives: to find resilience by creating the conditions for more community. Since then we have gathered family, friends (old and new), and neighbors around that table. I have my own conversational question prompts I like to use for these dinners, but we have also utilized various card decks such as the *Let's Get Deep: Friends Edition* and *Live Deeply's Meaningful Relationships Collection* playfully placed under dishes to invite deeper conversations.

Another work of brilliant scholarship that has been required reading in my ritual studies courses over the last few years is Priya Parker's *The Art of Gathering*. Parker is all about intentional curation of events, no matter whether it is a large party or a dinner gathering around one table because "connection doesn't happen on its own. You have to design your gatherings for the kinds of connections you want to create." She offers so many wonderful insights that include things that most of us never think of, such as this one about how to end a gathering intentionally: "Closings are a moment of power. How you end your time together shapes your guests' experience, sense of meaning, and memory of the event."

I suppose I'm simply making the case that if we are more intentional about the design of even our most common gatherings, we can curate more occasions of awe that heighten our sense of connection. We can create storytelling gatherings, especially with family elders, around a fire pit–one of the earliest ways our humanoid ancestors created awe. We can get together and sing–whether that's around a piano singing holiday songs badly or joining a choir (another way my spouse and I have upped our connection and effervescent quotient this year). We can create a regular walking group and end each walk with a two-minute dance jam–walking in time together is one of the easiest ways to get the oxytocin flowing. The possibilities abound.

Our block of houses in urban Kansas City will turn 100 this year and so we are coordinating a Longest Table gathering in which our street will be lined down the middle with tables end-to-end in a grand potluck celebration. The visual of everyone around the "same" table will evoke a sense of connection that we don't usually experience behind our closed doors or in the smaller gatherings of neighbors that hang out together. We'll find photos from city archives, explore what life was like for these folks a century ago (including the painful history of segregation that is a part of the story) and, most importantly, make sure everyone knows they are never alone. Awesome.

Awe Practice: Synch Up

For almost twenty years, I held retreats at Lake Tahoe for spiritual leaders. The place itself created a container of awe. We ate, talked, sang, laughed, learned and ritualized together resulting in a palpable collective effervescence. But my favorite evening about mid-week was when I would lead a "drum circle" experience with a variety of percussion instruments. At the beginning many, if not most, were leaning back with skepticism about their ability to join in the rhythm ("I don't have any"). But when everyone started playing together, you could see even the most tentative ones start to jam when they realized they would never be left alone to keep the rhythm on their own.

We can be more confident when given the opportunity to "synch up." It doesn't have to involve a drum circle or a mass singing event. Simply the basic rhythm and dance of passing the deviled eggs at a dinner party helps even the most introverted get in touch with our innate ability (and need) to connect. This awe practice is the curation of connection. What will it be? Who can you collaborate with to make it happen?

Do not feel lonely, the entire universe is inside you. Stop acting so small. You are the universe in ecstatic motion. Set your life on fire. Seek those who fan your flames. – Rumi

Self-Giving

Witnessing acts of self-giving can create an awe response in us–what Dacher Keltner calls "moral beauty." Additionally, engaging in acts of kindness, compassion, generosity, and collaboration, can also create awe in us as the reward and pleasure centers in our brains are fired off. It is what has been called the "helper's high." So whether as an observer or an actor, the awe quotient involved in self-giving is research-tested and anecdotes abound.

Some of the benefit of this kind of awe is that it can pull us out of depression and anxiety, distracting us from our obsession with inward-looking loops of past failures and future worries. It "gets us out of ourselves" and we become more pro-social beings. An experience during a transitional time of my life helped me take on the motto: "If you don't know what to do with your own life, do something for others."

After many years touring with the professional dance companies, I decided to make a change for my mental health. I was so very privileged to have been a part of something that very few dancers get the chance to do, but it is also a grueling and sometimes toxic world.

I was at the height of my dancing ability and so was not ready to hang up the dancing shoes (only a metaphor in the world of modern dance where bare feet are the norm). But I needed a change. I found myself helping to start a dance company in Naples, Italy, for a few months. At the end of that contract, I wasn't sure what to do or where my life was going. I made a bus trip to the English-speaking church in Rome one Sunday and met a group of children who had come to sing at the church. They were from Naples and they lived in an orphanage. Spirit simply plopped a purpose right in my lap. I approached the director and asked if they needed help. They explained they couldn't pay me but would take whatever volunteer help I could give. I did have work on the horizon but not for a few months. And so that was the beginning of living for a time onsite as a "mother" to ten little boys, ages ten and eleven. I had never had the urge to raise children but this began a time in my life in which I discovered the joy of being "mother-adjacent."

The awe I experienced at Casa Materna (Motherly Home) was totally that feedback loop: I witnessed with awe the resilience of these orphans and also felt the benefits of the "helpers high" to my psyche as I worked and played alongside them. I was both flooded with the gift of their presence and experienced a sense of purpose that changed an ambiguous period into a soul-nurturing time of my life. I can remember one night when a super-moon sat on the horizon of the Gulf of Naples and the reflection of light seemed to span the water, creating a path right up to where I was standing on the beach. I felt deep trust that a path would be illuminated for my life. I was not alone. That image has returned to me again and again when I need that reminder.

"Experiencing awe prompts people to help others," reports Adam Hoffman, in his article, "How Awe Makes Us Generous," but it takes a shift in our sense of self. Dacher Kelter describes the "default self" as one that is distinct, independent, in control, and seeks to prevail over others. However, after awe-inducing events, Keltner's experiments showed that people had a cognitive shift into what he calls the "small self" in which "our individual self gives way to the boundary-dissolving sense of being part of something much larger." One well-known example in our modern imaginations is the "overview effect" reported by astronauts seeing the earth from space.

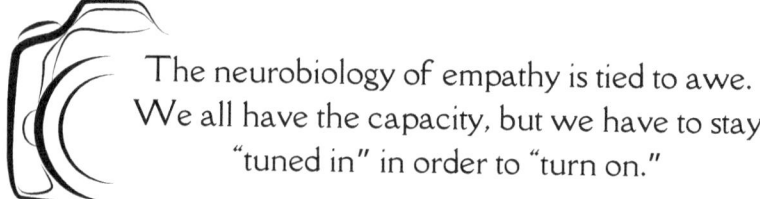

The neurobiology of empathy is tied to awe. We all have the capacity, but we have to stay "tuned in" in order to "turn on."

We don't have to become an astronaut to experience this shift. Any kind of awe will do it, Keltner says. Coming into contact with nature, from grand vistas to tiny details as well as simply watching videos of nature or funny animals are reported to improve generosity and a willingness to collaborate for good. Just hearing stories about extraordinary acts of kindness has a viral quality that inspires and engages our human abilities for mimicry based on the activity of mirror neurons. Collective effervescence elicits feelings of the small self as we physiologically bond with others and sense the truth of our interdependence. The studies are many and the experiments vary, but they all point to the phenomenon that when we are exposed to things that evoke the "small self," we are motivated to exercise our best cooperative selves.

> This is the day YHWH has made—
> let us celebrate with joy!
> Please, YHWH, please save us!
> Please, YHWH, give us prosperity now!
> Blessings on the one who comes in the name of YHWH!
> We bless you from YHWH's temple!
> YHWH is God and God has enlightened us.
> Join the festal procession! With palm fronds in hand
> go up to the horns of the altar!
> – Psalm 118 (Inclusive Bible)

The Hebrew poet composed a song that would be used again and again throughout history, perhaps most notably at the entrance of Jesus on a donkey at one gate of the city of Jerusalem while Herod entered at another gate on horses with a display of "power-over." It is an iconic scene, like others in various religious narratives, that exemplify two ways of existing in the world: one that exclaims "let all join the festal procession of life" and one that threatens those who need to "know their place" at the margins, never invited to join, much less lead, the parade. Marches in the name of justice continue as this human saga of inequality drones on.

The default self's penchant for individualism and "getting ahead" is not all bad as these attributes continue to aid our survival as we work to attain goals of advancement. But if we are not equally committed to ethical development alongside these advances, I'm not sure we can really call it progress. The attributes of the default self were perhaps evolutionarily helpful but as they have grown to toxic proportions it seems they are stunting the evolution of our empathic abilities–our ability to see and respond to the whole planet and the wellbeing of all. At my most cynical I wonder if we can escape what feels like a design flaw in the human species. But I have found hope in this research and the possibility that if we claim our role as Purveyors of Awe in our lives and communities we can somehow be part of a movement to help humanity regain the empathic ground we've lost.

What is an example from your life of feeling a cognitive shift from the "default self" to the "small self?" What concerns do you have about the trajectory of humanity and what helps you to experience hope and resilience?

O Divine Sovereign, grant that I may not seek to be consoled, as to console. To be understood, as to understand. To be loved, as to love. For it is in giving that we receive. It is in pardoning that we are pardoned. It is in dying that we are born to eternal life. – St. Francis of Assisi

A dear friend and mentor of mine, Dave Wendleton, was my pastor when I was a youth. He had such an impact on my experience of curating occasions of awe and spiritual depth. Our youth group did things like hold Easter vigils in a candlelit room of the church and marched down Main Street of our little rural town with Dave blasting his trumpet on Easter morning. He went on to become a leader in Clinical Pastoral Education (where pastors learn to serve in hospital settings) and later hospice work. His love of awe has extended to a love of the mountains where he now lives and his beautiful singing bowls that he uses in meditation. On a recent visit, he shared with me two stories from his career walking alongside those who are dying. One involved a former park ranger who yearned to be in the forest one last time. Arrangements were made for EMTs to take him on a stretcher to the middle of a grove of trees. Wide-eyed wonder gave way to a peaceful death right then and there, in the midst of this man's beloved sacred space. In another instance, a nurse asked Dave to stay with a woman who was close to the end of her life in a hospice room until she got back. The nurse returned with a horse and led it right into the woman's room through an outside patio door. It turns out this patient had a long history with horses and as the horse nuzzled her, she drifted peacefully out of this life realm.

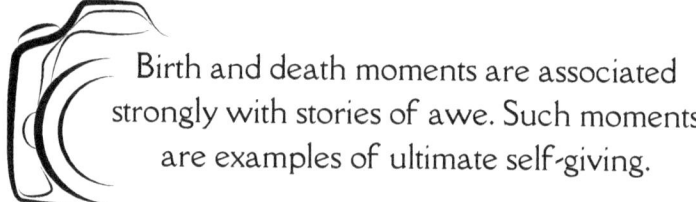

Birth and death moments are associated strongly with stories of awe. Such moments are examples of ultimate self-giving.

Dacher Keltner says that awe can lead to a greater understanding of the cycles of life and death. He talks about the amazing neurophysiology of the release of feel-good chemicals in the brain between parents and babies in the first six months. These exchanges mirror the same physical constructs of experiences of awe. Studies show that the continued exposure to wonder contributes to the wellbeing of children as they grow. Throughout our lives and perhaps even in our deaths, as we see in Dave's stories, awe can help us into more peaceful states. Optimum levels of respiratory sinus arrhythmia (RSA) occur when the calming of the parasympathetic system signals the vagus nerve to send messages to the heart to speed up on inhalation and slow down on exhalation, resulting in a state of relaxation and peace. Perhaps for the dying, relaxed states of awe assist in letting go into ultimate peace.

Keltner's own work on awe, as I've mentioned before, was piqued by the death of his brother. He describes the awe he felt when abiding with him during those last hours of life, aided by the practices of knowing, bearing witness, and compassionate action he had learned from Roshi Joan Halifax, a teacher and author of *Being with Dying*. Afterwards, grief threw him into a time of feeling "aweless." Eventually though, his study of the stories reported by the thousands of people in his research helped him open to the residual presence of his brother in "gentle breezes and in being embraced by a powerful, warming sun." He has come to believe that our awe companions in life remain with us in "ever more mysterious ways after they leave, enabling an opening to new wonders of life." Seeking awe is where we can tune into these moments of continued connection.

In 1935 Albert Einstein said, "The fairest thing we can experience is the mysterious. It is the fundamental emotion which stands at the cradle of true art and true science. He who knows it not and can no longer wonder, no longer feel amazement, is as good as dead, a snuffed-out candle." Death can happen in many ways, including a kind of walking-dead state of numbness to the awe that exists both in extraordinary events and in the everyday, in the ordinary, all around us. There is much today that seeks to deaden our ability to thrive. When basic needs are not only disregarded but destroyed by policies that bolster the already-privileged to the demise of the vulnerable, when fear maxes out our nervous systems such that the greatest percentage of doctor's visits are related to mental health, when we are duped into believing that "protecting our own" is the only way to survive, the freedom to wonder, to delight, and to connect in meaningful ways is harmed.

But Einstein's "cradle" where artists and scientists work is alive and well, as we have seen, and they urge us to awaken and to heal through practices of awe that, despite the movement to extinguish our candles, can renew our love for the life we've been given and spur us with energy for making this world a better place. The elements of curating a life of spiritual depth–beauty, wonder, meaning, curiosity, delight, connection, and self-giving–are essential to sustaining our ability to work for the very survival of our species and all species who inhabit this planet, as well as the planet itself that we call home. We must become Purveyors of Awe, practicing for ourselves and encouraging others. We must curate conditions for wellbeing by adding to our caring repertoire not only the basics of sustenance for the physical bodies of our fellow humans, but also the capacity for awe that keeps our candles lit.

Awe Practice: Light the Candle for Awe

By now you know my obsession with candles. Part of it has to do with the moving light they create that induces instant awe for me, but also because various sources of light have been used not only practically, but symbolically, by humans throughout our history and across places and spiritual traditions. The Sunday after September 11, 2001, the church I was attending sought to offer some way to respond in a time when words felt wholly inadequate. So we filled a table with candles and invited people to come light them during an extended time accompanied by meditative music. This action was so simple and yet so profound because it was a ritual action that could contain whatever people needed to express at the moment–grief for those lost and experiencing loss, outrage, concern for those who would be wrongly targeted by hate, and prayers for peace. It was so powerful that it became a weekly ritual long after people forgot its origin. So...

Find a candle that can be a reminder for you about whatever aspects of awe-seeking have become resonant for you. Light it at whatever intervals is possible for you. Set it somewhere in your living space where you will see it and be reminded that we need your light, your aliveness, in this world.

Don't ask what the world needs. Ask what makes you come alive, and go do it. Because what the world needs is people who have come alive.
— Howard Thurman

Dear Spiritual Leaders...

I'm looking at you... clergy of all faiths, counselors, therapists, chaplains, spiritual directors, social workers, spiritual entrepreneurs, activists, somatic practitioners, travel guides, filmmakers, artists, curators, teachers, professors, child care workers, senior living activity directors, among others. My net of what comprises "spiritual leaders" is wide because I know there are so many who care deeply for people's spiritual wellbeing–tied as it is to emotional, mental, physical, and social flourishing.

AND I'm looking at you, spiritual communities... faith communities, affinity groups of all kinds, neighborhoods, support and recovery groups, friends groups–essentially any gathering of people who care for each other's welfare and care about the extended human family.

This guided study and journal is but a small dip into the world of how the elements of spiritual depth grounded in practices of awe can contribute to the project of the survival and wellbeing of humanity. I specifically called it *Purveyors of Awe* because I'm hoping you will take on the mantle of heightening the instances of awe practice in your particular fields of influence. So many are wondering in this moment what they can do in the face of problems that seem insurmountable. There are many answers to that question and wonderful organizations that are helping us engage in the alleviation of suffering, from essential feeding, clothing, and sheltering needs to equally necessary ways to keep people safe from harm. As I studied the research on awe, it became clear to me that not just *after* we get things sorted out, but *as* we work on the never-ending list of threats to flourishing, the opportunity exists to curate these awe-filled elements that can help us not just *stay* alive, but *be* alive.

Whether it is a therapist's prescription for a period of wandering in an awe-inspiring place such as the one I was given, organizing a meditative walk in the woods for a support group, providing space for an digital detox gathering in your church, or inviting a drum facilitator to lead your colleagues in some radical connectivity through collective effervescence, my plea to you is "just do it." Whether you know for a fact it will contribute to a little or a lot of healing for what ails us, go for it. Try it. That's the great thing about being a Purveyor of Awe, it doesn't have to take a lot of effort, just intention and the passion to give it a go. People will resonate differently with some forms of practice than others (and I hope you are understanding that there are so very many more ideas than I have presented here).

In the next few pages, I'll offer some ideas for "A Year of Awe" that communities might consider sponsoring and examples of those who are already doing this. You will come up with other ideas based on your context and your experiences. I just want to get your imagination stirred up. What if once a month your town, neighborhood, or group knew there was an opportunity to engage these elements in delightful and hospitable ways? Churches and other communities of faith, I want to specifically say to you: I'm not talking about events exclusively curated only for your community, within your four walls. I'm suggesting more of what I call "casual encounters of the spiritual kind" that are less tied to an invitation to worship on Sunday morning. Spiritual leaders and communities, I challenge you to see yourselves more as an established group of people who simply have the gifts of hospitality, organization, and a passion for wellbeing deep in your DNA.

I'm so deeply grateful for all of you in this moment. Not only for the work you do, but who you are in this world. And I also hope that you see how much we need your wellbeing and spiritual depth. I have been blessed, as you have seen in these pages, with people who guided me, lifted me, encouraged me, healed me when, as an artist and spiritual leader, I experienced the burnout that is "for real" in our line of work. Just writing this to you is part of the story of my own wellbeing. Thank you.

Peace & Passion,
Dr. Marcia McFee

A Year of Awe

Here are examples of events to curate in community perhaps on a monthly basis over the course of a year. A "year" is totally random and there are not an even twelve suggestions here and they do not appear in any suggestion of an order. These ideas are meant to spark your own ideas for spreading the good vibes where you are based on your context and capabilities. If you are reading this in print form, scan the QR code at the end of this section to get to an online version with active links. To see more ideas and share your own, use that QR code or go to this link to our Padlet idea board.

Unplugged Event: The Offline Club
The Offline Club is an organization that was founded in the Netherlands as a counter-cultural movement to seek intentional respite time from our always-online, hyper-connected digital world. You could offer an offline retreat day during which members of your community are invited to spend some real face-to-face time together. There could be supplies for activities like journaling, art-making, crafting, going for a walk, playing games, or anything else that gets you away from your screens for a time. When you check out the Offline Club, notice that they curate the time with equal parts individual time and group interaction.

Candlelight Music Event
You may have heard about the popular candlelight music concerts that offer up a sensory-rich experience of music by the warm glow of candlelight. Here's a small-scale house concert clip to give you a sense of the atmosphere. You could stage your own candlelight concert experience! The use of battery-operated candles rather than real flames can help to ensure the safety of your musicians and attendees as well as the comfort of scent-sensitive people. Keep an eye out for good sales on these kinds of candles so you can increase their numbers over time (and start a battery fund!). I went to one of these events at a church in which the piano was placed in the center of an open space and we all laid on yoga mats or other kinds of pallets in a sun-ray configuration emanating out from the center. Awesome!

Group Singing Event

The Gaia Music Collective is a vocal workshop group based in NYC that offers experiences for up to 200 singers to form a one-day choir and learn to sing a three-part arrangement of a popular song. Check out their awesome videos on TikTok and YouTube. This could be a lot of fun to offer to the wider local community! Choir directors are uniquely equipped to provide leadership for this type of experience. Most people don't have time to commit to weekly choir at church or in community choirs. But there are likely plenty of people of various singing ability who want to gather just once in a while for a "big sing" and some amazing collective effervescence! Choose something meaningful, perhaps connect it to a cause, and prepare the time for optimum fun and success. See the links to the social singing initiative *Koolulam* that I featured in this resource in the References section below.

Percussion Circle

Another wonderful idea for collective effervescence is to host a drum circle. I call it a "percussion circle" because having a range of percussion from drums to woodblocks, tambourines, and little homemade rattles made of old prescription bottles filled with popcorn is super fun and inviting for those who might be intimidated by a larger drum. This is an accessible way to invite many people into the experience of keeping time, making music, and perhaps even dancing together. I studied drum circle facilitation with Arthur Hull whose techniques for making drum circles really fun and interesting (instead of just banging away, which gets really monotonous) can be found in a series of lessons and videos here. If you don't have a couple of steady rhythm-keepers in your group, play along to a variety of percussion tracks from the amazing array of "world music" available!

Trip to a Botanical Garden or Zoo

Spending time in nature and with animals is a wonderful portal to experiencing awe and wonder. Your local zoo or garden may even offer events like this where people can engage in an intentional journey of meditation and contemplation. Of course, you could put together your own guided journey! Divide the time between individual calm observational practices with perhaps contemplative music-in-the-headphones options and delightful surprises that can be facilitated with a fun scavenger-hunt type activity. Remember that for those that can't get out as easily, you can curate a "trip to the zoo" or "trip to nature" by putting together videos and music, making the experience super-accessible. My friend and collaborator, Chuck Bell, has a series of YouTube videos of nature with his orchestral compositions available here.

Awe Walk or Forest Bath
In a similar vein, the practice of forest bathing or taking an awe walk can be a powerful way for participants to experience nature and achieve a heightened state of earth awareness that can be done anywhere, anytime. Create opportunities to do this in community and/or create a one-page instruction sheet or audio track to guide people whenever they have the opportunity to dive in.

Street Wisdom Walkshop
Simply taking a walk around your neighborhood or other spot in your city can be an eye-opening experience when you pay close attention to your surroundings. Street Wisdom is a collective that teaches how to offer and experience "walkshops" designed to help us get in touch with our environment and form deep connections with our local community. Hear founder David Pearl describe the impetus for establishing this organization, and listen to his TEDx talk about how Street Wisdom cultivates practices of attention, intention, and connection.

The Longest Table
Connecting as a community over a meal is nothing new, but you might consider offering a new shape to this experience by hosting a Longest Table event which I described in the Connection chapter. This can be a powerful way to illustrate the depth and breadth of your community in a visual way, and create new opportunities for connection between friends and neighbors.

The People's Supper
Gathering around the table to engage in honest and challenging conversations is one important way we can grow as a community. Here is a guide from a collective called The People's Supper about how to establish "brave spaces" and lead deep conversations, especially in "purple" communities.

The Table Prayer Interfaith Gathering
I collaborated with my mentor, friend, and hymn writer extraordinaire Ruth Duck, and dear friend and frequent collaborator, the beloved and prolific musician Mark Miller, on an evening of eating, singing, blessing, and conversing in an interfaith gathering. It was published by Hope Publishing and is available here for free download. If you are a faith leader, organize an interfaith choir and collaborate with other spiritual leaders in your community for this night of connection.

Cooperative Game Night
Sometimes the best games are ones with no winners and losers. Consider hosting a cooperative game night with the goal of helping your community members get to know one another! One fun idea is cooperative Scrabble – Mattel released an updated version of the classic game called Scrabble Together that offers more collaborative and less competitive rules for gameplay (originally conceived by Quakers!). Alas, the licensed product is only available in the UK as of this writing, but there are plenty of "house rules" variations on classic Scrabble that you could try!

365 Days (or a Month) of Wonder
Deborah Farmer Kris is the author of a new book on parenting that prioritizes creating experiences of awe for children (release date May 2025). Check out her suggestions for engaging in an intentional practice of noticing and sharing about our experiences with awe. You could encourage people to send in photos, make comments on a social media thread, tag your organiazation in their personal posts, or any number of ways to encourage engagement. See more about her work in the References section below.

Labyrinth
If you have access to some open space (use your own indoor or outdoor space or get a permit to do this in a city park, a parking lot, or a green space), you can create your own labyrinth and invite people to use it as a walking meditation experience (get out those battery-operated candles and line the journey). Watch an inspirational video about an urban organization called the "Racial Healing Hub" that created a community-participation labyrinth here. See many more examples and ideas on our Pinterest board that is all about labyrinths (my favorite is the labyrinth made of canned goods to donate to the food bank). Or create a smaller version such as the example of my New Year's Eve ritual in the Meaning section of this resource.

Hygge Workshop

As I wrote about in the Delight section of this resource, *Hygge* is a Danish and Norwegian term that refers to a sense of coziness and contentment that inspires curating spaces and occasions that are comforting to the senses. Check out this description of Hygge Workshops that could inspire you to create your own (candle-making, knitting, and woodworking are fun activities related to *hygge*).

Visit a Donkey Rescue Farm or Start a Pet Ministry

The donkey farm from the Delight section hopefully has inspired you to check if similar animal sanctuaries exist around where you are. I highly recommend supporting these kinds of organizations that offer awe through respite for ailing nervous systems by planning a visit, volunteering, and donating funds. A new book about "pet ministry" is forthcoming from Upper Room Books but in the meantime, explore this connection to animals and wellbeing at the author's Pinnacle View United Methodist Church page here.

Scan the QR code to access a digital version of this section.

References

These are the references used throughout this resource, in order of appearance. I highly recommend any of the books or articles for further study. If you are reading a print version of this resource, use the QR code at the end of this section to get to a digital version of these Reference pages to access the links in this list.

All scripture references from *The Inclusive Bible: The First Egalitarian Translation*. Lanham, MD: Rowman & Littlefield Publishers, 2007.

Introduction

Haskell, David, and Godspell Ensemble. "Prepare Ye the Way of the Lord." Track 1. *Godspell (Original Off-Broadway Cast Recording)*. Bell Records and Arista/BMG Records, 1971, vinyl.

Keltner, Dacher. *Awe: The New Science of Everyday Wonder and How It Can Transform Your Life*. New York: Penguin Press, 2022.

Weill, Simone. "Attention and Will" in *Gravity and Grace*. First English edition published in Abingdon, Oxon and New York: Routledge, 1952.

Beauty

Farley, Wendy. *Beguiled by Beauty: Cultivating a Life of Contemplation and Compassion*. Louisville, KY: Westminster John Knox Press, 2020.

O'Donohue, John. *Beauty: The Invisible Embrace.* New York: HarperCollins Publishers, 2004.

"Why We Should Seek Beauty in the Everyday Life (The Science of Happiness Podcast)." Greater Good Magazine. The Greater Good Science Center at the University of California, Berkeley, 2024.

Wonder

Wesley, Charles. "Love Divine, All Love's Excelling." Public Domain, 1747.

Gardner, Howard. *Frames of Mind: The Theory of Multiple Intelligences*. New York: Basic Books, 1983.

Parker, Monica C. *The Power of Wonder: The Extraordinary Emotion That Will Change the Way Your Live, Learn, and Lead*. New York: Penguin Random House, 2023.

Interplay founded by Cynthia Winton-Henry and Phil Porter.

Meaning

Wagner, Jane. *The Search for Signs of Intelligent Life in the Universe*. New York: HarperCollins, 1986.

McLaren, Brian. *Life after Doom: Wisdom and Courage for a World Falling Apart*. New York: St. Martin's Publishing Group, 2024.

Dave Brubeck Jazz Quartet and the Murray Louis Dance Company. "Take Five."

Curiosity

Shigeoka, Scott. *Seek: How Curiosity Can Transform Your Life and Change the World*. New York: Hachette Book Group, 2023.

Farmer Kris, Deborah. *Raising Awe-Seekers: How the Science of Wonder Helps Our Kids Thrive*. Free Spirit Publishing, 2025.

Rick Steves's website: https://www.ricksteves.com

Percy, Walker. *The Moviegoer*. New York: Farrar, Straus, and Giroux, 1960.

Delight

Jason Silva's YouTube channel: https://www.youtube.com/@ShotsOfAwe

Nikolai / Louis Foundation for Dance: https://www.nikolaislouis.org

Axel, Gabriel, dir. *Babette's Feast*. 1987; Denmark: Nordisk Film, 2001. DVD.

Zen Donkey Farms website: https://zendonkeyfarms.com/pages/meet-the-herd-version-2

Wiking, Meik. *The Little Book of Hygge: Danish Secrets to Happy Living*. New York: HarperCollins Publishers, 2017.

Connection

Paul Winter Consort: https://paulwinter.com/paul-winter-consort

McNeill, William H. *Keeping Together in Time: Dance and Drill in Human History*. Cambridge MA: Harvard University Press, 1995.

Durkheim, Émile. *Les Formes élémentaires de la vie religieuse: Le système totémique en Australie*. Paris: F. Alcan, 1912.

Koolulam website: https://www.koolulam.com

Lehrman Bloch, Karen. "The Sacred Light of Koolulam." Jewish Journal. Tribe Media Corporation, September 21, 2022.

Gaia Music Collective YouTube channel: https://www.youtube.com/@gaiamusiccollective

Brown, Brené. "Why Experiencing Joy and Pain in a Group Is So Powerful." Greater Good Magazine. The Greater Good Science Center at the University of California, Berkeley, 2024.

77

Redmond, Layne. *When the Drummers Were Women: A Spiritual History of Rhythm*. New York: Three Rivers Press, 1997.

Let's Get Deep: Friends Edition card game

Live Deeply's Meaningful Relationships Collection cards

Parker, Priya. *The Art of Gathering: How We Meet and Why It Matters*. New York: Riverhead Books, 2018.

Self-Giving

Hoffman, Adam. "How Awe Makes Us Generous." Greater Good Magazine. The Greater Good Science Center at the University of California, Berkeley, 2024.

Rudd, Melanie, Kathleen D. Vohs, and Jennifer Aaker. "Awe Expands People's Perception of Time, Alters Decision Making, and Enhances Well-Being." *Psychological Science* 23(10). 2012. 1130–1136. DOI: 10.1177/0956797612438731

Halifax, Joan. *Being with Dying: Cultivating Compassion and Fearlessness in the Presence of Death*. Boston, MA: Shambhala Publications, 2009.

Einstein, Albert. *The World as I See It*. Translated by A. Harris. London: John Lane The Bodley Head, 1935.

Scan the QR code to access a digital version of this section.

Author

Dr. Marcia McFee is a retreat leader, professor, worship designer, author, keynote speaker, and ritual artist. Drawing on a first career in professional dance and musical theater and equipped with a Master's in Theology from Saint Paul School of Theology and a PhD in Liturgical Studies and Ethics from the Graduate Theological Union, she has taught for over thirty years, exploring mindsets and skillsets for creating extraordinary portals through which communities journey with the Spirit. The task is at once deeply theological and philosophical, extraordinarily tied to modern discoveries that are increasing our appreciation of the world, and wonderfully artistic. She has traveled extensively as a consultant, keynote speaker, and leader of worship, having designed and led worship for regional, national, and international gatherings of several denominations. She is the creator and visionary of the Worship Design Studio (www.worshipdesignstudio.com), an online experience of coaching, education, inspiration that has served thousands of churches for over fifteen years. She is the author of *The The Worship Workshop,* a workbook for worship teams and, as an avid skier, wrote *Spiritual Adventures in the Snow: Skiing and Snowboarding as Renewal for Your Soul*. Her third book, *Think Like a Filmmaker: Sensory-Rich Worship for Unforgettable Messages*, has become a best-seller and is utilized by seminaries and churches all over the world. Purveyors of Awe is her current project which will include interviews, retreats, and keynotes.

You can see more about the accompanying worship series already produced for churches at www.worshipdesignstudio.com/awe (Lent version and Anytime version).

Contact Dr. McFee for retreats and keynotes at marcia@marciamcfee.com.

www.ingramcontent.com/pod-product-compliance
Lightning Source LLC
Chambersburg PA
CBHW050444010526
44118CB00013B/1678